AQA

AS/A LEVEL

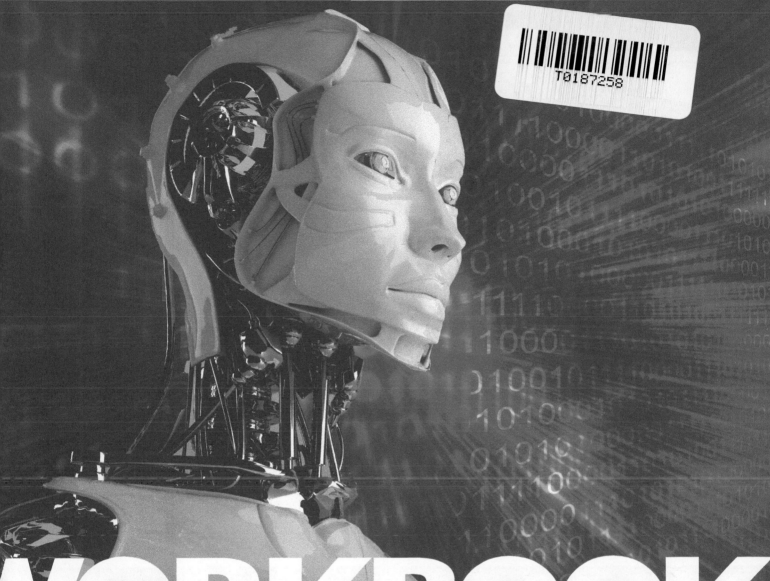

WORKBOOK

Computer Science 2

Mark Clarkson

HODDER
EDUCATION
LEARN MORE

Contents

1 This workbook will help you to prepare for the AQA Computer Science Paper 2 exam.

2 Your exam is 2 hours 30 minutes long (1 hour 30 minutes for AS Level) and includes a range of questions. You should answer every question on your exam paper.

3 For each topic of Paper 2 there are:
- stimulus materials including key terms and concepts
- short-answer questions that build up to exam-style questions
- space for you to write or plan your answers.

4 You still need to read your textbook and refer to your revision guides and lesson notes.

5 Answering the questions will help you to build your skills and meet the assessment objectives AO1 (knowledge and understanding), AO2 (application) and AO3 (design, program and evaluate computer systems).

6 Marks available are indicated for all questions so that you can gauge the level of detail required in your answers.

7 Timings are given for the exam-style questions to make your practice as realistic as possible.

8 Answers are available at:
www.hoddereducation.co.uk/workbookanswers

Fundamentals of data representation

Number systems, number bases and units of information

In order to understand how data is represented in computer systems it is essential to start with representing numbers.

Number systems

Different types of variables are used depending on the sets of numbers they may need to store. You are expected to be aware of how to describe sets of numbers using their name, symbol and example sets, e.g. $\mathbb{N}$, the set of natural numbers, where $\mathbb{N} = \{0,1,2,3,\ldots\}$.

Number bases

You are expected to be able to convert between numbers represented in binary (base 2), denary (base 10) and hexadecimal (base 16). It is also useful to be able to explain why different number bases are used in different situations.

Units of information

While a binary digit (or bit) is used to store a single 0 or 1, much larger units of information are used to describe file sizes, storage capacities and data transmission capacities. You are expected to be familiar with both data units in powers of 10 (kilobytes, megabytes, gigabytes, terabytes) and data units in powers of 2 (kibibytes, mebibytes, gibibytes and tebibytes).

---

1 a **Give examples from the options below for each of the following sets. (AO1)** *7 marks*

- –3 - 7 - 0 - $\sqrt{-3}$ - π - 6.3 - $\sqrt{3}$

  i **a natural number**

  ....

  ii **a member of set $\mathbb{Z}$ that is not a member of set $\mathbb{N}$**

  ....

  iii **a member of set $\mathbb{Q}$ that is not an integer**

  ....

  iv **an irrational number**

  ....

  v **a number that is not in the set $\mathbb{R}$.**

  ....

b **Define what is meant by a rational number.**

  ....

c **Explain why all integers can be described as rational numbers.**

  ....

② Numbers can be represented using binary, denary or hexadecimal number bases.
(AO1, AO2)

9 marks

a Complete the table below by converting each positive integer into its alternative representations.

Binary	Denary	Hexadecimal
1011		
	90	
		37
1100 1101		

b Explain why computer scientists often prefer to use hexadecimal representation instead of working with binary numbers.

....

③ State the maximum number of unique values that can be stored using each of the following. (AO1, AO2)

3 marks

a 3 bits

....

b 1 nibble

....

c 1 byte

....

④ State the value, in bytes, of the following units. (AO1)

4 marks

a 8 bits

....

b 1 kB

....

c 1 KiB

....

d 5.3 MB

....

Binary number system

All numbers in a computer system are ultimately stored using binary. It is important to recognise and to be able to represent binary numbers using different rules.

Unsigned integers arithmetic

The simplest representation only deals with whole numbers and is relatively straightforward to convert. Unsigned binary is also the only form that you will be expected to add or multiply.

Two's complement integers

In two's complement the left-most/most significant bit (MSB) is used to indicate the sign (positive or negative) and the remaining bits signify the value.

Fixed point fractional numbers

As the value of each bit doubles as you move left, the value also halves as each bit moves right, and so 0.11 in binary represents $0 + \frac{1}{2} + \frac{1}{4}$ (or 0.75).

Floating point fractional numbers

Floating point works in a similar way to standard form. The binary number is made up of a mantissa and an exponent, so 3.25 can be represented as 0.1101 | 001 where 01101 is the mantissa (value) and 001 is the exponent (amount to shift the bits). Shifting the bits by 2 (the exponent) we get 11.01, or 3.25.

5. Complete the table to show the positive and negative representation for each integer. (AO1, AO2) **6 marks**

Denary number	Positive binary	Negative binary
37	0010 0101	1101 1011
73		
	0111 1100	
		1001 1100

6. Add each of the following pairs of binary numbers, showing your working and where you have carried numbers. (AO1, AO2) **4 marks**

   a **0101 + 1010**

   ....

   ....

   ....

   b **1011 0101 + 0111 1101**

   ....

   ....

   ....

   c State the name of the problem encountered in your answer to part (b).

   ....

7. Multiply the following pair of binary numbers, showing your working. (AO1, AO2) **3 marks**

   0010 0110 * 0000 0011

   ....

   ....

   ....

8. Convert the fractional binary numbers into their denary equivalent and the denary numbers into their binary equivalent. (AO1, AO2) **6 marks**

Binary representation	Denary representation
0101 . 1000	
1001 . 0110	
1011 0111 . 1111	
	11.75
	9.625
	147.3125

9. Convert each of the following two's complement, floating point binary numbers into their denary equivalent. (AO1, AO2) **4 marks**

   a Mantissa: 0.111 1101   Exponent: 0101

   ....

   ....

   b Mantissa 1.100 1101   Exponent: 0111

   ....

   ....

10. Convert each of the following denary numbers into their normalised floating point representation. Write each answer with an 8-bit mantissa and 4-bit exponent. (AO1, AO2) **6 marks**

   a −82

   ....

   ....

   b 0.0625

   ....

   ....

11. For each pair of numbers, the first shows the original number and the second shows the value of the closest binary approximation. For each one calculate the absolute error and the relative error (as a percentage, to two decimal places). (AO1, AO2) **4 marks**

   a Actual value: 4.6   Stored value: 4.5625

   ....

   ....

   b Actual value: 17.1   Stored value: 17.125

   ....

   ....

Information coding systems

Information coding refers to character sets (ASCII and Unicode) and error checking and correction.

12. The ASCII code for the letter 'A' is 65. The ASCII code for the letter 'a' is 97. Complete the table below to show the ASCII codes for a variety of letters. (AO1, AO2) **6 marks**

Character	ASCII – upper case	ASCII – lower case
A	65	97
B		
D		
	76	

13 ASCII is a 7-bit character set. Another character set uses 16 bits. (AO1, AO2)   **4 marks**

a State the maximum number of unique characters that could be stored using ASCII.

....

b State the maximum number of unique characters that could be stored using a 16-bit character set.

....

c Suggest two reasons why a larger character set might be chosen for a given application.

....

....

14 A keyboard is used to transmit characters to a computer system using ASCII, in which the digit 0 is represented by 48_{10}. Even parity is used with the most significant bit (the left-most bit) being used as the parity bit. (AO1, AO2)   **5 marks**

a State the binary code that is transmitted when the user presses the key for the digit 0.

....

....

b Another key is typed and the bit pattern 1011 0011 is received. State whether the computer system will accept or reject this code and why.

....

....

c State two reasons why parity bits are only partially effective at identifying transmission errors.

....

....

....

15 A 4-bit binary message is transmitted from one computer system to another using majority voting. The following code is received: 001 111 110 000. (AO1, AO2)   **9 marks**

a State the original binary message.

....

b Explain how majority voting is used to detect and correct errors in data transmission.

....

....

....

c Describe how a check digit could be used to detect and correct errors in data transmission instead.

....

....

....

d Explain the difference between a checksum and a check digit.

....

....

Representing images, sound and other data

Images and sounds are also stored as bit patterns, used to represent analogue/continuous data using a digital/discrete format. Analogue to digital convertors (ADCs) and digital to analogue convertors (DACs) are used to capture analogue inputs and recreate analogue signals.

Digital representation of images

Images are most frequently stored as bitmaps, in which each pixel is used to represent a small block of colour. Image file size and quality can be optimised by altering the pixel dimensions, resolution and colour depth of an image. Digital image files typically include metadata, i.e. data about the data.

It is also important to consider vector graphics, made up of points, lines, shapes and colours and their relative advantages and disadvantages to bitmap graphics.

Digital representation of sound

Analogue sounds can be sampled in order to be stored digitally, whether through recording sound using a microphone or directly through the electrical output from a guitar, for example. Nyquist's theorem is essential in determining the minimum sample rate for capturing audio. Similar to image files, sound files can be compressed in terms of sample rate (equivalent to resolution) and sample bit depth (equivalent to colour depth). Just as graphics can also be represented as vectors, so too sounds can be stored using MIDI in which the pitch, timbre, length and volume of each note is stored.

Compression

As well as the lossy compression methods described above, both run length encoding (RLE) and dictionary-based methods of lossless compression exist, each with positives and negatives in comparison to lossy methods.

16 An image is stored as a bitmap graphic with pixel dimensions of 3000 px wide by 2000 px tall. Each pixel is represented by an 8-digit binary number. (AO1, AO2)   `7 marks`

   a   State the resolution of the image as a single number.

   b   State the maximum number of unique colours that can be represented in an 8-bit image.

   c   Excluding any metadata or compression, calculate the size of the image file.

   d   Suggest three pieces of metadata that may be stored in the file.

17 A bitmap image measure 420 px wide by 100 px tall and is made up of 12 different colours. (AO1, AO2)   `3 marks`

   a   State the minimum colour depth for the image.

   b   Calculate the file size of the image in bytes, excluding metadata and compression.

18 A student is planning a piece of digital art work and is unsure about whether to use a vector or bitmap format. (AO1, AO2) `5 marks`

    a   State three pieces of data that might be stored in a vector graphic file.

    ....

    ....

    ....

    b   Describe one common use of vector graphics and one common use for bitmap graphics.

    ....

    ....

19 A researcher is recording a 30-second audio file that will be played to dogs, who can hear sounds up to 60 kHz. (AO1, AO2) `6 marks`

    a   State the minimum sample rate that should be used for the audio file. Justify your answer.

    ....

    ....

    b   The researcher chooses to record the audio at a sample rate of 100 kHz, with a bit depth of 12 bits per sample. Calculate the file size of the recording.

    ....

    ....

    ....

    c   Describe the impact on the quality and file size of the audio recording for choosing a bit depth of 24 bits per sample.

    ....

    ....

20 A musician is composing a new piece of music and is unsure whether to store it as a MIDI file or a sampled audio file. (AO1, AO2) `3 marks`

    State three items of data that would be stored about each note in a MIDI file.

    ....

    ....

    ....

21 Two sections of a DNA sequence are shown below. (AO1, AO2) `6 marks`

TTTTTTTTTAAAA   CGTAACGTAACGT

    a   Describe, with an example, how run length encoding (RLE) could be used to reduce the file size needed to store the first sequence.

    ....

    ....

b   Describe how a dictionary could be used to reduce the file size needed to store the second sequence.

....

....

....

Encryption

Only two algorithms for encryption are named in the specification and you are expected to be familiar with both. While a Caesar cipher is one of the simplest algorithms, it is very weak. The Vernam cipher is considered perfect in terms of security but is almost impossible to use in the real world. Most other ciphers lie somewhere in between the two and include ciphers that are considered computationally secure in that they cannot be cracked quickly or easily, though they are able to be broken given sufficient ciphertext and time.

22 Figure 1 shows a cipher wheel used to encrypt simple text messages. The outer wheel contains the plaintext and the inner wheel contains the ciphertext equivalent.
(AO1, AO2)

The plaintext message 'DAY' would be encrypted as 'GDB'.

Figure 1

a   Encrypt the plaintext 'ELEPHANT' using the Caesar cipher with the settings shown in Figure 1.

....

b   Decrypt the ciphertext 'DSULO' using the Caesar cipher with the settings shown in Figure 1.

....

c   The Caesar cipher shown in Figure 1 has been used with a key of –3. Encrypt the plaintext 'ABBA' using the Caesar cipher with a key of –5.

....

d   Encrypt the plaintext 'ABBA' using the Caesar cipher with a key of +21.

....

e   Explain the significance of your answer to part (d).

....

f   Describe, with an example, one possible approach to cracking the Caesar cipher when given a ciphertext message but no key.

....

....

....

23 The table shows the binary representation, using 8-bit ASCII, for the first five letters of the alphabet. (AO1, AO2)

8 marks

Letter	ASCII
A	0100 0001
B	0100 0010
C	0100 0011
D	0100 0100
E	0100 0101

a   Show the result of applying an XOR to the binary messages '0100 0011' and '1100 1010'.

....

....

....

....

b   Using the key '1010 0111 0101 1011 1001 0110', encrypt the message 'BAD' using the Vernam cipher, showing your working.

....

....

....

....

c   State three requirements for the Vernam cipher key that are required to keep the plaintext message completely secure.

....

....

....

d   The Vernam cipher is an example of symmetric key encryption. Explain what is meant by the term 'symmetric key'.

....

....

....

Exam-style questions

24 In the table below a tick is placed in a box to indicate that the number shown is a member of a particular set. A number can be a member of more than one set. (AO1, AO2)   `5 marks`

    a Complete the table by ticking one or more boxes in each row.

Number	$\mathbb{Z}$	$\mathbb{R}$	$\mathbb{Q}$	$\mathbb{N}$	Irrational
3.7					
−9					
4					
π					

    b Explain the purpose of ordinal numbers.

....

....

....

25 Convert each of the following denary numbers into 8-bit, two's complement binary. Show any intermediate steps in each case. (AO2)   `3 marks`

    a 17

....

....

....

    b −126

....

....

....

26 Each of the following numbers uses two's complement binary. Convert each number into its denary equivalent. Show any intermediate steps in each case. (AO2)   `3 marks`

    a 0011 1010

....

....

    b 1101 0000

....

....

27 The following value is stored in a byte: (AO2)   `4 marks`

1	0	1	0	1	1	0	0

    a If the value represents an unsigned binary number, state its value in denary.

....

....

b   If the value represents an unsigned binary number, state its value in hexadecimal.

....

....

c   If the value represents a two's complement binary number, state its value in denary.

....

....

d   If the value represents an unsigned, fixed point binary number with four bits to the left of the decimal point and four bits to the right of the decimal point, state its value in denary.

....

....

28 Below is a value that is stored as a two's complement, floating point binary number with an 8-bit mantissa and a 4-bit exponent. (AO1, AO2)   `9 marks`

Mantissa: 0.001 0101   Exponent: 0101

a   State the value of the number in denary. You must show your working.

....

....

b   Explain what is meant by a normalised floating point number and state the normalised equivalent of the binary number shown above.

....

....

c   Write the normalised floating point representation of 25.75 as a binary number with an 8-bit mantissa and a 4-bit exponent.

....

....

d   The closest approximation to 33.68 that can be stored using an 8-bit mantissa and a 4-bit exponent is 33.75. Calculate the absolute error that has occurred.

....

....

....

e   Calculate the relative error that has occurred, as a percentage to two decimal places.

....

....

29 A message needs to be sent from one device to another. (AO1, AO2)   `12 marks`

a   Describe one advantage of sending the message using Unicode rather than ASCII.

....

....

....

b The message is to be sent using ASCII, but is also going to be encrypted. State what is meant by encryption.

....

....

....

c A test message 'ABC' is to be sent using the Vernam cipher. The ASCII code for 'A' is 0100 0001 and the key being used is 0101 1010 0010 0110 0111 1101. Encrypt the message 'ABC' using the Vernam cipher with the key provided. Show your working.

....

....

....

d Explain why the Vernam cipher is considered more secure than the Caesar cipher.

Write your answer on a separate sheet of paper.

30 A digital image measures 6000 px wide by 4000 px tall and uses a colour depth of 24 bits per pixel. (AO1, AO2)   `12 marks`

a State what is meant by a pixel.

....

....

....

b Calculate the total file size for the image file. Show your working.

....

....

....

c State two items of metadata that could be included as part of the image file.

....

....

d Describe two possible methods of lossy compression that could be applied to the image.

....

....

....

e Explain how run length encoding could be used to reduce the file size of the image.

....

b ....

....

Fundamentals of computer systems

Hardware and software

Hardware can be defined as a physical device that forms part of a computer system and is dealt with in Section 3. Software can be defined as the programs that run on the computer system.

Software classification

Software can be broken up into systems software (software to enable the computer system to run) and application software (software that allows the user to do something). Systems software can be further broken down into four categories: operating systems, utilities, libraries and translators.

Operating systems

The basic role of an operating system is to hide the complexity of the hardware from the user. More specifically, the operating system is responsible for managing various elements of the computer including peripherals, processors, memory and users as well as providing a user interface.

---

1 Phillipa has an operating system installed on her computer. (AO1)   `5 marks`

a   **State the purpose of an operating system.**

....

b   **State four different types of management carried out by the operating system.**

....

....

....

....

2 Below is a list of software classifications. (AO1)   `10 marks`

- Utility programs
- Library programs
- Application software
- Translators
- Operating systems

a   **For each of the following examples, state which of the above would be the most appropriate software classification.**

i   Compiler

....

ii   Disk defragmenter

....

iii   Image editor

....

    iv   **IDE**

....

  b   **Explain what is meant by a library program.**

....

....

  c   **State the name and describe the purpose of two other utility programs.**

....

....

....

....

Classification of programming languages and types of translator

You are expected to be aware of two classifications of programming languages. These are low-level languages (machine code and assembly language) and high-level languages (e.g. VB, Java, C++, Pascal, Python, etc.).

Low-level languages

Machine code refers to the binary instructions passed directly to the processor. Assembly language uses mnemonics (short, more memorable instructions using acronyms or shortened words) to represent each instruction. The two types of languages are both platform specific (so a machine code program written for an iOS device would not work on an Android device) and instructions map one to one, so a program with 120 lines of assembly language code would be translated to 120 lines of machine code. Translation is carried out using a type of translator called an assembler.

High-level languages

High-level languages such as the one you will be using in Paper 1 are typically not platform specific (i.e. the finished program can be translated and run on different types of computer system) and one line of high-level code can often be translated into many lines of low-level code. One subset of high-level languages is imperative high-level languages. In an imperative language, a sequence of instructions is carried out in a programmer-specified order. The programmer describes how to solve the problem. This is different from some other examples of high-level languages, including SQL and functional programming, which are dealt with later in this book. High-level languages can be translated using a compiler or an interpreter, each of which has its own positives and negatives.

3 Jamal has written a computer program using assembly language and has used a translator to turn his program code into machine code. (AO1, AO2)   3 marks

  a   **Describe one similarity and one difference between the assembly language program code and the machine code.**

....

....

b State the name of the translator used to convert the assembly language code into machine code.

....

4 Jemima has created a computer-based board game that is designed to work on desktop computers, tablet computers and smartphones. (AO1, AO2) `5 marks`

a Suggest whether the program should be written using a high-level language or a low-level language. Justify your response.

....

....

....

....

....

b Jemima is using a compiler to translate the program code into machine code. State the name of another suitable translator.

....

5 A program, has been written using an imperative high-level language. When translated the compiler produces bytecode. (AO1, AO2) `6 marks`

a Explain what is meant by the term 'imperative'.

....

....

....

b Explain the term 'bytecode' and describe the purpose of producing bytecode rather than machine code when the program is compiled.

....

....

....

....

....

....

Logic gates and Boolean algebra

Logic circuits can be represented in several different forms. These include logic circuit diagrams, truth tables and Boolean algebra.

Logic gates

There are six basic logic gates to be aware of: NOT, AND, OR, XOR, NAND and NOR.

Boolean algebra

When using or simplifying Boolean algebra there are three basic rules to remember:

Boolean identities

For example:

A.A = A AND A, which is equivalent to A.

A.$\bar{A}$ = A AND NOT A, which is impossible, simplifying to 0.

Factorise/expand brackets

Boolean algebra behaves exactly like normal algebra, thus:

A($\bar{A}$ + B) = A.$\bar{A}$ + A.B

This further simplifies (using a Boolean identity) to 0 + A.B

and then simplifies again to A.B

De Morgan's Laws

The simple way to remember this is that if you break or join a bar, you change the sign, thus:

$\overline{A.(A+B)} = \bar{A} + \overline{(A+B)}$

The double bar on the right will cancel, and so:

$\bar{A} + \overline{\overline{(A+B)}} = \bar{A} + A + B$

This further simplifies to 1 + B, and then again to 1.

By applying these three rules whereever possible, you will always be able to find a suitable solution.

---

6 **Figures 2 and 3 show two logic gates. (AO1, AO2)**   9 marks

Figure 2   Figure 3

a   **State the name of the logic gate in Figure 2.**

....

b   **Complete the truth table below for the gate shown in Figure 2.**

A	B	P
0	0	
0	1	

c   State the name of the logic gate in Figure 3.

....

d   Complete the truth table below for the gate shown in Figure 3.

A	B	P
0	0	
0	1	

e   State the name of the logic gate that is represented by the following truth table and draw its circuit diagram.

A	B	P
0	0	0
0	1	1
1	0	1
1	1	0

....

....

f   State the name of a logic gate that has only one input.

....

7 Figure 4 shows part of a logic circuit diagram. (AO1, AO2)   4 marks

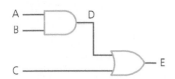

Figure 4

a   Complete the truth table below for Figure 4.

A	B	C	D	E
0	0	0		
0	1	1		
1	1	1		
1	0	0		

b   Describe the logic circuit shown in Figure 4 using a Boolean expression.

....

8 Simplify the following Boolean expressions. (AO2) 19 marks

a   1.A

b $X.\overline{X}$

c $0 + Y + \overline{Y}$

d $A.(\overline{A} + B)$

e $(P + Q).(\overline{P} + \overline{Q})$

f $\overline{(\overline{a} + b)}.a$

g $\overline{\overline{((x.y).x)}.\overline{y}}$

9 Some computer systems use a D-type flip-flop. (AO1)   5 marks

a State the purpose of a D-type flip-flop.

....

....

b Describe the two inputs into a D-type flip-flop.

....

....

....

....

Exam-style questions

60

10 There are four different types of systems software. (AO1)   17 marks

a Describe the purpose of systems software.

....

....

b One example of systems software is an operating system. Name and describe the function of the three other types, including examples.

Write your answer on a separate sheet of paper.

c Describe the role of a typical operating system and its key functions.

Write your answer on a separate sheet of paper.

11 Paul has written a program to collect data about train timetables and delays which is capable of automatically collecting published data and then carrying out analysis on that data. He wishes to share his program with others so that they can use and improve it. (AO1, AO2)   15 marks

a Explain the need for using a translator on the high-level source code and the steps involved in translation.

....

....

....

....

....

....

b Describe the advantages of writing code in a high-level language rather than machine code.

....

....

....

....

....

c Explain why it may be preferable to use an interpreter rather than a compiler to translate the high-level source code.

Write your answer on a separate sheet of paper.

12 Simplify the following Boolean expressions. (AO2)   12 marks

a B.0

....

b $r + \bar{r}$

....

c $a.b + a.\bar{b}$

....

....

d $\overline{(\bar{a} + b)}$

....

....

....

e Simplify $\overline{(a + b).\bar{a}}$ and $\overline{b.(\bar{a} + \bar{b})}$ to prove that they are equivalent.

Write your answer on a separate sheet of paper.

13 A computer process will begin as soon as one of Input A and Input B are on and as long as Input C is on and Input D is off. (AO1, AO2)   6 marks

The Boolean variable X should be set to TRUE if the values of the variables A, B, C and D indicate that process X can start and to FALSE if they indicate that process X cannot start yet.

a In the space below, draw a logic circuit that will represent the logic of the system described above for the inputs A, B, C and D and the output X.

A •

B •

C •   • X

D •

b   Describe the logic circuit shown in your previous answer as a Boolean expression.

....

....

....

14 The circuit diagram in Figure 5 shows part of a logic circuit. (AO1, AO2)   `7 marks`

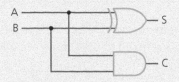

Figure 5

a   Complete the truth table below for Figure 5.

A	B	C	S
0	0		
0	1		
1	0		
1	1		

b   State the name of the logic circuit described in Figure 5.

....

c   Explain the purpose of the logic circuit described in Figure 5.

....

....

d   Explain the purpose of using two circuits, identical to that described in Figure 5, together and name the extra gate that would be required to make the circuit work.

....

....

....

15 Part of a circuit diagram is show in Figure 6. (AO1, AO2)   `3 marks`

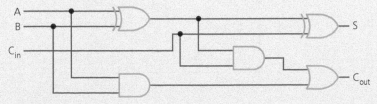

Figure 6

Complete the truth table below to show the output from Figure 6 when applied to the inputs provided. You must show your working.

A	B	C_{in}					
0	1	1					
1	1	1					
1	1	0					
1	0	0					

Fundamentals of computer organisation and architecture

Internal hardware components

The internal hardware of a computer system is made up of a processor and main memory, connected by three buses (address, data and control). Additionally, input and output (I/O) controllers are used to communicate with external devices including keyboards, monitors (VDUs) and secondary storage.

The stored program concept, structure of the processor and its components

Processor components

The processor itself contains an arithmetic and logic unit (ALU), control unit (CU), clock and registers, both general purpose and special purpose. The fetch–decode–execute (FDE) cycle describes how the components interact. Processor performance is affected by clock speed, cache size, number of cores and also word length and bus widths.

Processor instruction set

Each processor design uses its own instruction set. Each instruction is made up of an opcode (which details the operation to be carried out and the addressing mode) and an operand (the value or address to perform the operation on). With direct addressing, the operand refers to a memory location (the operation is directed to a value from memory). With immediate addressing, the operand refers to a specific value (available immediately).

External hardware devices

These include input and output devices including RFID scanners, barcode scanners, laser printers and digital cameras and also storage devices including magnetic, optical and solid state devices.

1. **Figure 7 shows part of the internal architecture of a computer system. (AO1, AO2)** `10 marks`

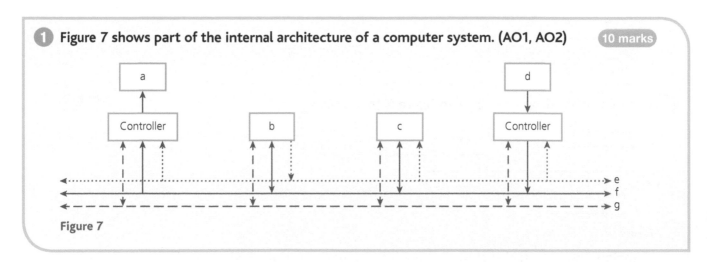

Figure 7

a   Complete the table below for the labels shown in Figure 7.

Label	Component name
a	
b	
c	
d	
e	
f	
g	

The address bus in the current system is made up of 16 lines. The data bus in the current system is 32 bits wide.

b   State the number of additional lines required in the address bus to double the number of available addresses.

....

c   Describe the impact of increasing the width of the data bus.

....

....

2   The table below describes the six steps in the fetch–decode–execute cycle, though not in the correct order. (AO1, AO2)

Execute	MBR ← [Memory] addressed	PC ← [PC] + 1
MAR ← [PC]	Decode	CIR ← [MBR]

a   Put the steps in the correct order and then explain what is happening at each stage, using the full name of each register.

Step number	Step	Description
1		
2a		
2b		
3		
4		
5		

b   Explain why the steps numbered 2a and 2b take place simultaneously.

....

....

....

c   Explain what happens when an interrupt occurs during the fetch–decode–execute cycle.

....

....

....

....

d   Describe the difference in design between the Harvard and von Neumann architectures.

....

....

....

e   Suggest one typical use for a computer system using Harvard architecture.

....

3   Andrew has purchased a new laptop package that includes an external DVD drive,
    HDD backup drive, an internal SSD and a laser printer. (AO1)   15 marks

    a   Explain the need for a secondary storage device in a computer system.

....

....

    b   State the name of an optical device that comes with Andrew's computer.

....

    c   State the name of a magnetic device that comes with Andrew's computer.

....

    d   Describe five advantages for using an internal SSD instead of an HDD in Andrew's laptop.

....

....

....

....

....

    e   Explain the principles of operation of a laser printer.

....

....

....

....

....

....

....

....

....

....

Generic assembly language instruction set

LDR R<d>, <memory ref>	Load the value stored in <memory ref> into Register <d>
STR R<d>, <memory ref>	Store the value stored in Register <d> into the address specified by <memory ref>
ADD R<d>, R<n>, <operand>	Add the values of <operand> and Register <n> and store the result in register <d>
SUB R<d>, R<n>, <operand>	Subtract the value of <operand> from the value of Register <n> and store the result in register <d>
MOV R<d> <operand>	Move the value of <operand> into Register <d>
CMP R<n>, <operand>	Compare the value of Register <n> and the value of <operand>. Used in conjunction with B <condition> <label>
B <label>	Always branch to the point in the program labelled with <label>
B <condition> <label>	Branch to the point in the program labelled with <label> if the condition is met. MUST ONLY BE USED IMMEDIATELY FOLLOWING CMP R<n>, <operand> Examples: BEQ – Branch if equal BNE – Branch if not equal BGT – Branch if greater than BLT – Branch if less than
AND R<d>, R<n>, <operand>	Perform a bitwise AND on Register <n> and the value of <operand> and store the result in Register <d>
ORR R<d>, R<n>, <operand>	Perform a bitwise OR on Register <n> and the value of <operand> and store the result in Register <d>
EOR R<d>, R<n>, <operand>	Perform a bitwise XOR on Register <n> and the value of <operand> and store the result in Register <d>
MVN R<d>, <operand>	Perform a bitwise NOT on the value of <operand> and store the result in Register <d>
LSL R<d>, R<n>, <operand>	Perform a logical shift left on Register <n> by the number of bits stored in <operand> and store the result in Register <d>
LSR R<d>, R<n>, <operand>	Perform a logical shift right on Register <n> by the number of bits stored in <operand> and store the result in Register <d>
HALT	End program

Note: <operand> **will be presented in the form:**

- **#3 (immediate addressing, the value 3)**
- **R3 (direct addressing, the contents of register 3)**

To answer this question you should refer to the generic assembly language instruction set above.

4 **The table shows the current state of the registers in a processor. (AO1, AO2, AO3)**   **14 marks**

Register	Current value
R0	0110 0100
R1	0000 0010
R2	0000 0000

a **State the result of carrying out the instruction MOV R2, #73. Only write in the value for the register that has changed.**

Register	Current value
R0	
R1	
R2	

b   State whether the operand #73 uses direct or immediate addressing.

....

c   State the result of carrying out the instruction AND R0, R0, R1. Only write in the value for the register that has changed.

Register	Current value
R0	
R1	
R2	

d   State the purpose of the previously executed instruction.

....

e   Write an assembly language program to carry out the following high-level algorithm.

```
B ← A // 4
B ← B * 4
IF B = A THEN
  C ← 1
ELSE
  C ← 0
```

The mathematical operator // represents an integer division.

Assume that register R1 holds the value of A, R2 holds the value of B, R3 holds the value of C and registers R4–R10 are available if needed.

Write your answer on a separate sheet of paper.

Exam-style questions

(30)

To answer this question you should refer to the generic assembly language instruction set on page 27.

5   Below is an assembly language program along with a section of main memory.
(AO1, AO2)

15 marks

Memory address	Value (denary)
600	3
601	12
602	9
603	14
604	11

```
    MOV R0, 0
    LDR R1, 600
    MOV R2, #601
labelA:
    CMP R1, #0
    BEQ labelB
    LDR R3, R2
    ADD R0, R0, R3
    SUB R1, R1, #1
    ADD R2, R2, #1
    B labelA
labelB:
    STR R0, R2
    HALT
```

a Complete the trace table below, in decimal, to show how the values stored in the registers and in main memory change as the program is run. You may not need to use all of the rows.

Registers				Memory				
R0	R1	R2	R3	600	601	602	603	604
				3	12	9	14	11

b Explain what the assembly language program above does.

....

....

....

....

c Describe four possible alterations to the design of a processor that could lead to faster execution.

....

....

....

....

6 A professional photographer has purchased a new computer with an internal magnetic hard drive, internal solid state drive, an optical drive and an external magnetic hard drive. (AO1, AO2)

8 marks

a Provide two reasons why the photographer has chosen a magnetic hard drive as an external device rather than a solid-state drive.

....

....

....

....

b The photographer wishes to send a collection of 300 wedding photographs to a client, using physical media. Compare and contrast the advantages of sending the images on a DVD, USB stick or portable magnetic hard drive, making it clear which you think is most appropriate.

Write your answer on a separate sheet of paper.

Consequences of uses for computing

This topic refers to the moral (effects on the individual), ethical (effects on society), legal and cultural consequences of developments in computer science. A broad range of questions could conceivably be asked, but a common theme is the responsibility of developers to consider the impact of both software and hardware platforms.

The best advice is to read the question very carefully in order to identify exactly what you are being asked and tackle this in a structured format.

1. A student has decided to develop an online lending library for digital content. Users will need to sign up with their name, email address and postal address (for verification) and will then be able to access one selected album, film or TV program for 24 hours. At the end of the 24 hours the user will be able to select a new digital product.

   State the names of two relevant pieces of legislation the student should be aware of and explain the relevance of each. (AO1).

   `6 marks`

   ....

   ....

   ....

   ....

   ....

   ....

Exam-style questions

`20`

2. In 2012, a charity offering pregnancy advice and pregnancy terminations was maliciously accessed and the names, addresses and phone numbers of several thousand people were copied. The organisation later admitted that they did not realise that the data was stored as part of their website and did not take steps to protect the data.

   In the context of this example, discuss:

   - how it was possible for this data to be stored as part of the website
   - what steps the owners of the charity could have taken to prevent the data from being accessed
   - what legal and ethical issues might have arisen as a result of storing this data as part of the website
   - what lessons the charity might have learnt from the incident and how their practices might have changed as a result.

   In your answer you will be assessed on your ability to follow a line of reasoning to produce a coherent, relevant and structured response. (AO1)

   `12 marks`

   Write your answer on a separate sheet of paper.

Fundamentals of communication and networking

Communication

The 'fundamentals of communication' explains some of the terminology and the basic principles behind different methods of transmitting data.

Parallel vs serial

In parallel transmission data, several bits are sent down several parallel wires or tracks at the same time. This has the advantage of being able to send more data at the same time, though there are overheads and potential data loss due to interference. Serial transmission sends the data one bit after another, reducing those issues, and is now the typical format used over any significant distance (e.g. USB, SATA).

Synchronous vs asynchronous

In synchronous transmission, a timing signal is required to ensure that each state change occurs at the same time. In parallel transmission, one line can be dedicated to this timing signal, but in serial transmission this is more difficult. In asynchronous transmission, start bits can be used to synchronise the sender and receiver.

Measuring transmission speed and capacity

It is important to understand the distinctions between bit rate, baud rate, bandwidth and latency – words which are often confused in general conversation.

Protocols

A protocol is an agreed set of rules (e.g. whether to use odd or even parity, synchronous or asynchronous transmission, the meaning of each potential signal within the available bandwidth). For A Level, it is necessary to be aware of a wide range of protocols (see Topic 4).

---

1 Since the year 2000, hard drive data communication has changed from favouring the PATA (Parallel ATA) protocol to the SATA (Serial ATA) protocol. (AO1, AO2)   **6 marks**

   a   Explain what is meant by the term 'protocol'.

   ....

   ....

   b   Describe the difference in operation between parallel and serial data transmission.

   ....

   ....

   ....

   c   State three disadvantages for using parallel data transmission for peripheral devices.

   ....

   ....

   ....

2 A computer network lists different values for the baud rate and bit rate of its transmission speeds. (AO1, AO2)

3 marks

   a   **State one other measure of network speed.**

....

....

   b   **Explain the impact of bandwidth on the bit rate.**

....

....

....

Networking

The networking topic is split into three sections: wired topologies, client–server vs peer-to-peer networks and wireless.

Topologies

The only two topologies that must be studied are the bus and star networks. The physical topology describes the physical layout of the network, but logically a star network can behave as if it were a bus network.

Networking between hosts

Many networks operate on a client–server model, including web servers, file servers, print servers and email servers. Clients make requests which are then fulfilled by the centralised server. In peer-to-peer networks, all hosts are equal and both data and processing can be distributed.

Wireless networking

The benefits of WiFi over wired networking are well known; however, collisions and security are significant areas to consider. Security can be improved in three ways.

1   Hiding the SSID of a network so that it cannot be found by those who don't already know it is there.

2   Encrypting the data sent across the network using WPA2 or similar methods.

3   Using MAC address filtering to only allow authorised devices to access the data.

Collision avoidance

To avoid collisions in wireless networking, CSMA/CA (carrier sense multiple access with collision avoidance) is used. With several devices capable of transmitting data, each device will check to see if there is a current broadcast before sending its own data. If two devices begin transmitting simultaneously then both will stop and wait a random time before trying again.

In addition, if RTS/CTS is also enabled then the sending devices will initially send a request-to-send message and will wait for a clear-to-send response before transmitting.

3 A firm of solicitors currently uses a local area network within their office. A central server holds details of clients, cases and appointments. The physical topology of the network is shown in Figure 8. (AO1, AO2)

8 marks

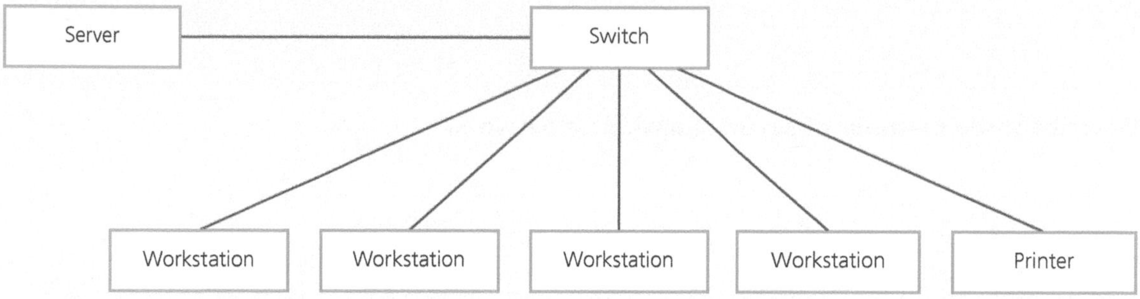

Figure 8

a State the name of the physical topology used in this network.

....

....

b Explain why the logical topology of the network may be different to the physical topology.

....

....

....

c Describe the role of the server in the network shown in Figure 8.

....

....

....

d A peer-to-peer network would remove the need for a server. Explain why this would not be an effective solution for the firm of solicitors.

....

....

....

....

4 A café currently has a small wired network consisting of a two PCs for office work, a modem and a printer. The owner of the café is considering the option of adding wireless functionality to the network. (AO1, AO2)

19 marks

a State the name of one piece of hardware that would be required to add wireless functionality to the network.

....

....

b Describe one benefit to the café of adding wireless functionality.

....

....

c   Explain why a wireless network may introduce new security risks to the network.

....

....

....

d   Describe three methods of securing a wireless network.

....

....

....

....

e   When a wireless device is ready to send a message, state three steps that the wireless device might carry out as part of CSMA/CA.

....

....

....

f   Explain one benefit and one drawback of using RTS/CTS.

....

....

....

....

The internet

The internet is the ultimate example of a wide area network (WAN) and allows devices across the globe to communicate. In order to do this, a large range of protocols is required. The key principles are described here and the individual protocols considered in more detail in Topic 4.

Packet switching

Routing and packet switching form the backbone of the internet. The ability to split large quantities of data into individual packets means that incorrectly received data can be corrected by resending only a few small packets instead of the entire transmission. It also means that individual packets can be routed across a different path so that bottlenecks can be avoided.

Addressing

Addressing on the internet is managed using uniform resource locators (URLs), IP addressing and domain name servers (DNSs). A URL allows for users to refer to a memorable address, such as files.hoddereducation.co.uk. A URL typically contains a host, a domain name and a top-level domain name. The top-level domain name indicates some information, such as the location or audience for the site (e.g. .gov, .uk, .biz). The domain name is designed to be recognisable by the user. The host indicates the individual server within the domain that is being addressed. Domain name servers provide a look-up table so that each domain name can be mapped to its IP address. Internet registries work together to ensure that two different clients cannot claim the same IP address or domain name, as each one must be unique.

Security

Common tools to improve the security of data include firewalls (designed to control what data can enter or leave a network), encryption (both symmetric and asymmetric), digital certificates and code monitoring to identify and protect from malware.

5 A student is using file transfer software to send data across the internet. (AO1, AO2) **15 marks**

  a Packet switching is used to transfer the data. Describe what is meant by 'packet switching'.

....

....

....

....

  b State three items of data that are included in a packet.

....

....

....

  c Explain the difference between a gateway and a router.

....

....

  d State one advantage of using a domain name instead of an IP address to identify the server.

....

....

  e Explain how DNS is used to allow the student to make use of the server's domain name.

....

....

....

....

  f State one situation where a DNS request will not be required when the student enters a domain name.

....

....

6 Jamal is responsible for configuring the security settings for a corporate network. (AO1, AO2) **18 marks**

  a Data can be encrypted using symmetric or asymmetric data encryption. State the difference between the two principles.

....

....

b Explain how private–public key encryption can be used to ensure that data can only be decrypted by the receiver.

....

....

....

....

c Private–public key encryption can also be used to authenticate a message using a digital signature. Describe the principles of operation of a digital signature.

....

....

....

....

....

d Explain the purpose of a digital certificate.

....

....

....

e Describe the principles of operation of each of the following types of malware.

  i   virus

....

....

....

  ii   worm

....

....

....

  iii   Trojan

....

....

....

The transmission control protocol/internet protocol (TCP/IP)

This topic is very content heavy and includes information on technologies that make use of the TCP/IP stack.

TCP/IP stack

For the purposes of this specification it is only necessary to be familiar with the four-layer TCP/IP stack (application, transport, network, link). Other models exist, including five- and seven-layer variants, however these will not be assessed.

Sockets

A socket is composed of an IP address and a port number. A device connected to a network must have an open socket in order to send or receive data, e.g. 192.168.0.1:80. Well-known ports include:

Port Number	Protocol
20 and 21	FTP
22	SSH
25	SMTP
80	HTTP
443	HTTPS
110	POP3

Protocols

A protocol is a set of rules or standards. It is expected that you are able to name and identify protocols relating to web pages (HTTP/HTTPS), email (SMTP/POP3), remote management (SSH) and file transfer (FTP) as well as the underlying principles for each type of communication.

IP addressing

As well as the basic structure of an IP address, both for IPv4 and IPv6, it is important to be aware that all IP addresses have a network identifier and a host identifier and that subnet masking is used to identify the length of the network identifier. In addition to this, you are expected to recognise the difference between public (routable) and private (non-routable) IP addresses as well as the principles of DHCP, NAT and port forwarding.

Client–server model

In terms of protocols the client–server model is awash with more acronyms. Once a websocket is opened, a CRUD application can use REST to carry out SQL functions and data can be transferred using JSON or XML.

---

7 Data is transmitted across a network using the TCP/IP protocol. (AO1, AO2) `12 marks`

   a   State the name and the purpose of each of the layers of the TCP/IP stack.

....

....

....

....

....

....

....

b Identify which two layers are required in order to open a socket.

....

c State the name of the layer that is used to store the MAC address of the recipient.

....

d Describe an advantage of separating the TCP/IP stack into individual layers.

....

8 A firewall is being configured and sockets need to be opened on some ports. (AO1, AO2)　**9 marks**

a Other than the port number, state what other information is needed to open a port.

....

b State one port number commonly associated with each of the following operations.

    i Sending and receiving unencrypted web pages

....

    ii Transferring files across a network

....

    iii Sending or receiving emails

....

    iv Remote management of a server

....

    v Sending and receiving encrypted web pages.

....

c In order to configure the firewall, the network administrator connects to the server remotely. Name the most suitable protocol to achieve this.

....

d Describe one advantage of connecting to the server remotely.

....

....

9 Figure 9 shows part of the physical topology of a local area network. (AO1, AO2)　**12 marks**

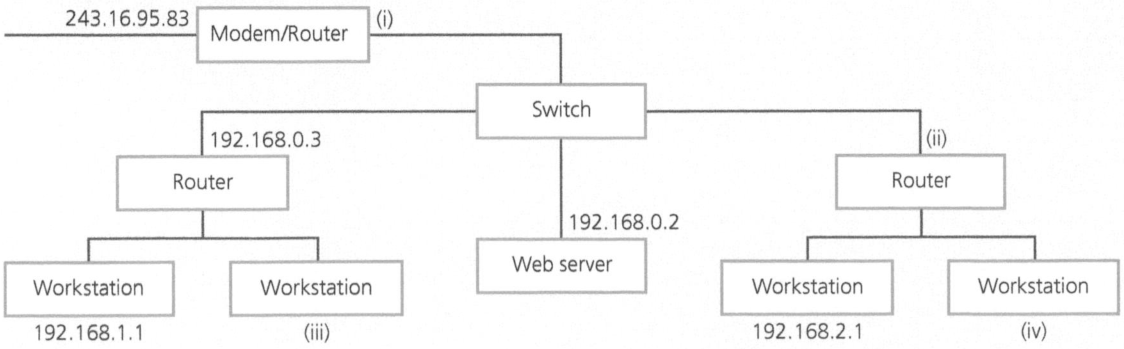

Figure 9

a State any IP address present in Figure 9 that is non-routable and explain the significance of this.

....

....

....

b State suitable IP addresses for (i), (ii), (iii) and (iv) in Figure 9.

....

....

....

....

c The IP address of the modem, 243.16.95.83, is composed of two identifiers. Using a subnet mask of 255.255.255.0 state the:

   i network identifier

....

   ii host identifier.

....

d The local area network contains over 1000 devices, but only has access to 245 public IP addresses. State why network address translation (NAT) is required for workstations to access the internet.

....

e Describe the process of using NAT within a pool of IP addresses.

....

....

10 An online database can be accessed using SQL at the command line or through a web interface. When accessed through the web interface, the websocket protocol is used. (AO1, AO2)   12 marks

a Explain the nature and purpose of the websocket protocol.

....

....

....

b State the meaning of the acronym CRUD.

....

....

c State the name of the four HTTP request methods described by REST and map each one to its relevant SQL command.

....

....

....

d State the name of two encoding mechanisms for returning the results of a query to a web page.

....

....

e State which of the two is preferred, being easier for humans to read and quicker for computers to parse.

....

Exam-style questions

11 A sensor is to be connected to a data logging computer system that is 3 metres away. (AO1, AO2)

5 marks

a State one advantage and one disadvantage of using parallel transmission.

....

....

b A decision is taken to use serial transmission. Explain why asynchronous data transmission must be used and why overhead must be added to the data.

....

....

....

....

12 A large company is using a local area network. Part of the network is shown in Figure 10. (AO1, AO2)

11 marks

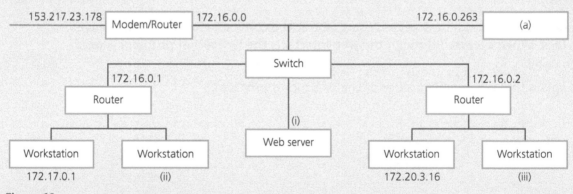

Figure 10

a Office staff have wireless access to the network. State the name of the device labelled (a) that is necessary to achieve this.

....

b The wireless network is only accessible to corporate staff. Describe two security strategies that will help prevent visitors from accessing the wireless network.

....

c State one IP address from Figure 10 that must be incorrect.

....

d State suitable IP addresses for (i), (ii) and (iii) in Figure 10.

....

....

e The network uses DHCP. Explain the function of using DHCP on the network, why it might be needed and one drawback.

....

....

....

13 An estate agent's website can be accessed by entering either 184.93.45.81 or www.barnabyestates.com into a web browser. (AO1, AO2)   `14 marks`

a State the domain name of the website.

....

b Explain why it is more common to use the URL for a website rather than the IP address.

....

c A company with the same name, in another country, wishes to use the same URL. Explain what measures are in place to prevent this.

....

....

Staff can access a database of existing and previous houses for sale using a webform to perform searches and to update the database.

d State which function of CRUD would be used to edit an existing record.

....

e Using REST, what HTTP method would map to an SQL function that would accomplish this?

....

f When receiving the results of the query, the data could be encoded using JSON or XML. Suggest two reasons why JSON may be preferable.

....

....

g The estate agency is considering replacing its existing thick-client infrastructure with thin-client technology. Discuss the advantages and disadvantages of replacing the IT infrastructure in terms of hardware requirements, capability and sustainability.

Write your answer on a separate sheet of paper.

Fundamentals of databases

Conceptual data models, relational databases, database design and normalisation

A database is a collection of tables, each relating to a single type of real-world entity, or relation. Each table is made up of fields which relate to attributes of that real-world object.

Primary and foreign keys

Each table should have a unique primary key, though composite keys can be used in which each combination of key values should be unique. A foreign key is where the primary key from another table is used to form a relationship between two tables. A database is likely to have multiple potential relationships.

Normalisation

Normalising databases means that duplication of data can be avoided and if data needs to be updated then each datum only needs to be changed once (e.g. if the channel number of a TV station changes). It is expected that students are able to normalise to the third normal form, which means that there are no repeating groups of attributes, that no data depends on a partial key and that there are no non-key dependencies.

---

1 Shown below is the design of a database used to store data about TV channels and their owners (e.g. BBC1 and BBC2 are both owned by the BBC. ITV1 and ITV2 are both owned by ITV). (AO1, AO2)

`7 marks`

```
Channel (OwnerID, ChannelID, ChannelNumber, Name)
Programme (ProgrammeID, Title, Genre, ChannelID)
Owner (OwnerID, Name, Country)
```

a State the name of one relation from the design.

....

b State the name of one attribute from the design.

....

c State the name(s) of the most suitable attribute(s) for a primary key in the Channel relation.

....

d State the name of an attribute which is acting as a foreign key and the name of the relation in which it has been used as a foreign key.

....

e On the incomplete entity-relationship diagram below, show the degree of any three relationships that exist between the entities.

```
┌──────────┐                    ┌──────────┐
│ Channel  │                    │ Programme│
└──────────┘                    └──────────┘



        ┌──────────┐
        │  Owner   │
        └──────────┘
```

2 A letting agent uses a database to keep track of its landlords along with the properties and tenants. A tenancy is agreed when a tenant has a contract to stay at a particular property. Historic records of tenancies are also stored in the database. (AO1, AO2) **8 marks**

The definitions of four relations in the database are shown below.

```
Landlord (LandlordID, FirstName, LastName, ContactNumber, Email)
Property (HouseNum, StreetName, Town, Postcode, PropertyType, LandlordID)
Tenant (FirstName, LastName, ContactNumber, Email)
Tenancy (TenancyID, TenantFirstName, TenantLastName, StartDate, Duration,
    MonthlyRent, SecurityBondAmt)
```

a The Tenant relation uses a composite key. Explain what is meant by a 'composite key' and why it has been used in this case.

....

....

....

It has been suggested that database is not fully normalised and that a new relation, Postcode, could be introduced to solve the problem.

b Explain what is meant by the term 'normalised'.

....

....

....

....

c Describe the changes that would need to be made to the database in order to incorporate the new relation. Use the same notation that has been used in the design above.

....

....

....

....

....

....

Structured query language (SQL)

Although Paper 2 is a written exam, you will be expected to compose your own statements as part of the database section. SQL can be split into two different types (actually four different types, though only two of them are part of the AQA specification).

Defining tables

Database definition language (DDL) is used to create tables. A DDL statement must include the attributes and their data types as well as the details of primary, composite and foreign keys. There are many, many valid data types and the mark schemes are lenient as long as suggestions are appropriate. You should at least make sure you are aware of INT, FLOAT, BOOL, CHAR, VARCHAR, DATE and TIME.

Manipulating data

Database manipulation language (DML) is used to perform four different types of operation, easily remembered using the acronym CRUD:

- **C**reate (add new data) using: INSERT INTO … VALUES (…)
- **R**etrieve (search for data) using: SELECT … FROM … WHERE
- **U**pdate (edit data) using: UPDATE … SET … WHERE
- **D**elete (remove data) using: DELETE FROM … WHERE

---

3 Shown below is the design of part of a database used to store the details of tests taken by A Level students (those in Year 12 or Year 13). Each student's ID is initially set to their surname and the first letter of their first name (e.g. Dave Jones would be given the student ID JonesD). If the student's name changes, their StudentID does not. **(AO1, AO2)**

10 marks

```
Student (StudentID, FirstName, LastName, YearGroup)
TestResult (ResultID, StudentID, TestID, DateofTest, Score)
Test (TestID, Topic, MaxScore)
```

a Complete the following DDL statement to create the Student relation, including the key field.

```
CREATE TABLE Student (
```

....

....

....

....

....

A new test with TestID 603 has been set. The test is about graph theory and has a maximum score of 80.

b Write the SQL commands that are required to record the test in the database.

....

....

The student with ID BloggsJ has recently moved from Year 13 into Year 12.

c Write the SQL commands that are required to record this change in the database.

....

....

....

The student with ID SmithD has recently left the college.

d Write the SQL commands that are required to remove the student's details in the database.

....

....

④ **Shown below is the design of part of a database used to manage an online audiobook subscription in which customers can 'purchase' one free audiobook per month as long as their subscription is still valid. Each book, purchase and author is identified using a unique numeric code. (AO1, AO2)** **15 marks**

```
Customer (Email, FirstName, LastName, EndDate)
Purchase (PurchaseID, CustomerEmail, BookID, Date)
Audiobook (bookID, Title, AuthorID, Reader, Length)
Author (AuthorID, First Name, LastName)
```

The length of the Audiobook *Alice in Wonderland* has been incorrectly entered as 1 hour 59 minutes when it should say 2 hours 59 minutes.

a   **Write the SQL commands that are required to record this change in the database.**

b   **Write the SQL commands that are required to find the title and reader for all of the books by Terry Pratchett.**

c   **Write the SQL commands that are required to find the title and author's full name for every book downloaded in June 2018, sorted in ascending order by the author's last name.**

Write your answer on a separate sheet of paper.

Big Data

Big Data refers to a collection of data that falls outside the normal type of everyday database. Most people understand this as being simply because the data set is too big (hence the name) but there are actually three main issues.

Volume

If there is too much data to fit on a single server then processing that data becomes problematic. The processing of the data needs to be distributed across several machines. The use of functional programming (Section 7) makes it significantly easier to write distributed code.

Velocity

A large data set can typically be associated with data that is being generated and collected continuously. YouTube users upload in excess of 300 hours of video per minute and at the time of writing, over 8000 tweets are sent every second. Data from GPS devices, sensors, cameras and audio devices such as Google Assistant and Alexa are continuously streamed and may need to be processed in real time.

(Statistics from www.fortunelords.com and www.internetlivestats.com)

Variety

The most significant challenge when processing Big Data is dealing with the nature of the data itself. Data is often unstructured and also comes in many different forms, including text, images and sounds. This means that a standard relational database structure is unsuitable as the data simply doesn't fit that format. Machine learning and pattern recognition are typical methods of tackling the challenges involved.

5 A data set has been classified as Big Data. (AO1, AO2)   6 marks

a Describe the three characteristics that might classify a data set as Big Data.

....

....

....

....

b In order to help process the data effectively, a distributed computer architecture is used. State
   the most suitable programming paradigm for handling Big Data in a distributed system.

....

c State two features supported by this programming paradigm that make it easier to process Big
   Data.

....

....

Exam-style questions

⏱ 40

6 Shown below is the design of part of a database used to manage the acts and stages
   at a music festival. Acts include music, comedy and a range of other entertainment types
   and may perform multiple times across the festival. Different stages have different curfew
   times dependent on the type of acts due to perform there and their location within the
   festival. Because the festival runs over several days, each booking includes the day of the
   week (Friday, Saturday or Sunday). TheBookingID and AgentID are made up of unique
   numeric codes. (AO1, AO2, AO3)   27 marks

```
Act (Name, ActType, Requirements, SetLength, Agent)
Stage (Name, StageType, Curfew)
Booking (BookingID, ActName, StageName, Day, StartTime)
Agent (AgentID, FirstName, LastName, Mobile, Email)
```

a Complete the following DDL statement to create the Booking relation, including the key field.

```
CREATE TABLE Booking (
```

....

....

....

....

....

....

....

b State the name of a primary key used within the database.

....

c State the name of a foreign key used within the database and name the relation in which it takes the role of a foreign key.

....

d The database is fully normalised. Explain what this means.

....

....

....

....

....

e On the incomplete entity-relationship diagram below, show the degree of any three relationships that exist between the entities.

| Agent | | Stage |

| Act | | Booking |

f A new stage, the 'Trapezoid' stage is to be added. This is to be a comedy stage and has a curfew of 11 p.m. Write the SQL commands that are required to record the new stage details in the database.

....

....

....

....

g In light of a recent health and safety review, all acts of type 'fire eater' are to be cancelled. Write the SQL commands that are required to remove their details from the database.

....

....

....

....

h The agent Paul Scott is threatening to cancel all bookings for his acts. Write the SQL commands that are required to find the act name, act type and the day of their scheduled performance. Display the results grouped by day, with Friday first and Sunday last.

....

....

....

....

....

....

....

....

....

There are restrictions on which acts can appear on which kind of stage. For example the:

- Trapezoid stage type is 'comedy' and only allows acts that have the type 'stand up' or 'comedy play'
- Little Top stage type is 'acoustic' and only allows acts that have the type 'folk' or 'classical'.

The information about which types of act can perform on which type of stage would be very helpful to the festival organisers.

i Explain how the database could be modified to represent which act types can perform on which stage type.

....

....

....

....

....

....

....

....

....

....

j If two members of the organising team attempt to edit a record at the same time then this could cause a problem for the integrity of the database. Explain how this problem could be prevented.

....

....

7 The diagram in Figure 11 shows part of the graph schema for an interactive map-based social media site in which people can share their current and previous locations. (AO1, AO2, AO3)

`12 marks`

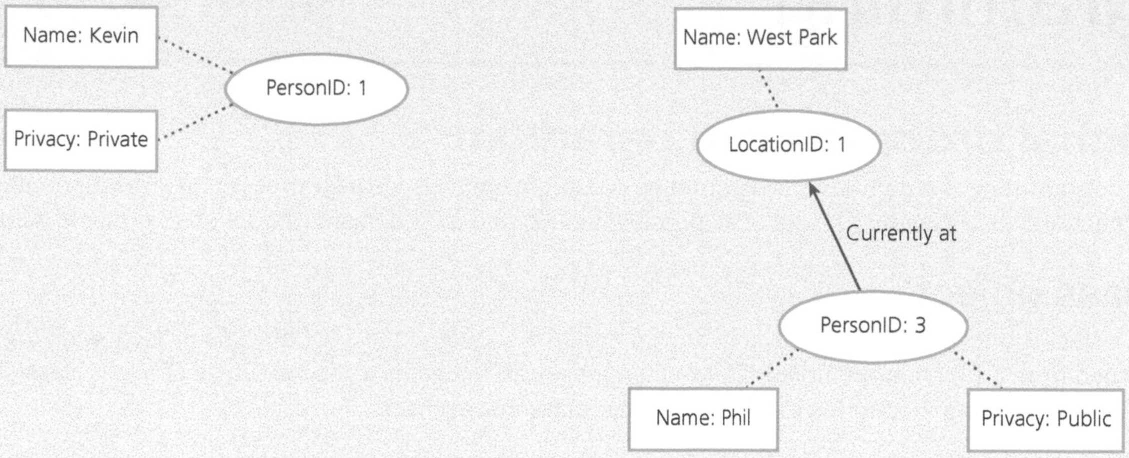

Figure 11

a   Complete the graph schema to represent the following:

- Kevin and Phil both 'follow' another user called Maryam.
- Maryam's privacy is currently set to public.
- Kevin has previously visited West Park.
- Kevin and Phil are brothers.
- Phil is a premium subscriber.

b   State the circumstances in which a large quantity of data might be sufficient to classify the data set as Big Data.

c   Provide two other aspects of the data set that could cause it to be classified as Big Data and explain why each one is problematic in terms of traditional data processing.

d   State what is meant by immutable data structures and explain why a programming paradigm that enforces the use of immutable data structures is preferable when processing Big Data.

Fundamentals of functional programming

Functional programming paradigm

Functional programming is a high-level programming paradigm which is ideal for processing large amounts of data. Each function has a domain (the set of all possible inputs) and a co-domain (the set of all possible outputs).

First-class object

A first-class object is one which can be used in an expression, be assigned to a variable, be passed as an argument and be returned from a function. In more high-level languages an integer is a good example of a first-class object. In functional programming, a function can be classed as a first-class object.

Function composition

A function, f(x) = x * 5.

Another function, f(y) = y + 4.

Applying x ∘ y means applying x of y, 5 * (y + 4).

Domain and co-domain

The domain of a function is the set of all possible inputs. The co-domain is the set of all possible outputs.

The function x takes an integer and squares it. This could be written as $f(x) = x^2$, or this could be written as f: A → B, where A is the domain (the set of all integers) and B is the co-domain (the set of all natural numbers, as the square of a negative number is always positive).

---

1 The sqrt function determines the square root of any positive integer, e.g. sqrt(25) = 5 or –5.

The dbl function calculates the double of any number that can be expressed as an integer or decimal fraction, e.g. dbl(3.2) = 6.4. (AO1, AO2)   `6 marks`

a   What is the domain of the function sqrt?

....

b   What is the co-domain of the function sqrt?

....

c   What is the domain of the function dbl?

....

d   What is the co-domain of the function dbl?

....

e   State the result of evaluating sqrt ∘ dbl(18).

....

....

f   State the result of evaluating dbl ∘ sqrt (a).

....

....

2 The times function takes an argument of two integers and returns their product,
e.g. times(2,6) = 12. (AO1, AO2)   **7 marks**

a   State the domain of the function times.

....

b   State the co-domain of the function times.

....

The partial application of times can be expressed in the form:

times: integer → (i) → (ii)

(Note that (i) and (ii) refer to missing elements.)

c   State what should be written in place of label (i).

....

d   State what should be written in place of label (ii).

....

e   Explain the purpose of partial application of a function, giving an example that uses the times
     function.

....

....

....

....

....

3 The following functions have been defined: (AO1, AO2)   **8 marks**

add: real x real → real

add(x, y) = x + y

sq: integer x integer → integer

sq(x) = x * x

A new function, sub, will take two integers as an argument and return the result of subtracting the
second number from the first number.

a   State the definition of the function sub.

....

....

....

b   State the co-domain of the function sub.

....

c State the result of evaluating add (7,(sq 3)). Show your working.

....

d State the result of evaluating sub (12,sq (4)). Show your working.

....

....

....

Writing functional programs and lists in functional programming

Head and tail

A list can be broken down into the head and the tail, so that a function can be applied recursively to the entire list.

For example, for the list L = [1,4,9,16], the head would be 1 and the tail would be [4,9,16].

If the list L = [12], then the head would be 12, and the tail would be [].

This is typically the base case in recursive functions.

When used in a function, the syntax f (x:xs) is used, with x being the head and xs being the tail.

High-order function

A high-order function is one which can take a function as an argument and/or can return a function. Map, filter and reduce/fold are examples of high-order functions.

For each of the following questions, the following functions and lists have been declared.

```
map f [] = []
map f (x:xs) = f(x) : map (f xs)
filter f[] = []
filter f(x:xs)  | f x = x: filter f xs | otherwise = filter f xs
fold f v [] = []
fold f v (x:xs) = f x (fold f v xs)
sq x = x * x
sqrt x = √x
dbl x = 2 * x
hlv x = x ÷ 2
a = [1,4,9,16,25]
b = [2,4,6]
```

4 State the result of evaluating each of the following functions. (AO1, AO2)   5 marks

a head(a)

....

b tail(a)

....

c   head (tail(tail a))

....

d   0:a

....

e   a ++ [36]

....

5 State the result of evaluating each of the following functions. (AO1, AO2)   **7 marks**

a   map sqrt a

....

b   map dbl a

....

c   filter >10 a

....

d   filter even a

....

e   fold + 0 a

....

f   fold * 0 b

....

g   fold * 1 b

....

6 State the result of evaluating each of the following functions. Show the result after
applying each function in turn. (AO1, AO2)   **7 marks**

a   map dbl (filter (<10) a)

....

....

....

....

b   fold (+) 10 (filter (>10) a)

....

....

....

....

c   fold (+) 0 (sq (filter (<10) a))

....

....

....

....

7 Answer the following questions. (AO2)   8 marks

a   The functions map, filter and fold are all high-order functions. Explain what is meant by a high-order function.

....

....

....

....

....

b   State the base case for the map function.

....

....

....

....

c   Explain the purpose of the fold function.

....

....

....

....

d   Explain how recursion is used to process the list b when evaluating the expression fold (+) 0 b.

....

....

....

....

c ....

....

Exam-style questions

(8) The following functions and lists have been declared. (AO1, AO2, AO3) `16 marks` ⟨20⟩

```
map f [] = []
map f(x:xs) = f(x) : map (f xs)
sq: integer → integer
sq x = x * x
dbl: real → real
dbl x = 2 * x
add: real * real → real
add x, y = x + y
a = [10,15,20]
b = [1,4,9]
c = [2,3,4]
```

a What is the co-domain of the function sq?

b Evaluate each of the following functions.

  i head a

  ii tail b

  iii b ++ c

  iv head a : tail b

c Evaluate each of the following functions.

  i dbl ∘ sq 3

  ii filter (even) a

  iii map dbl b

  iv fold (*) 1 c

  v fold (*) 2 b

d Write an expression that would result in the list [20,30,40].

e **Describe the recursive steps involved in evaluating the expression filter (<5) b.**

....

....

....

....

....

....

....

....

Cover photo: jim/Adobe Stock

Hachette UK's policy is to use papers that are natural, renewable and recyclable products and made from wood grown in well-managed forests and other controlled sources. The logging and manufacturing processes are expected to conform to the environmental regulations of the country of origin.

Orders: please contact Hachette UK Distribution, Hely Hutchinson Centre, Milton Road, Didcot, Oxfordshire, OX11 7HH

Telephone: 01235 827827

Email: education@hachette.co.uk

Lines are open from 9 a.m. to 5 p.m., Monday to Friday. You can also order through our website: www.hoddereducation.co.uk

ISBN: 978-1-5104-3702-9

© Mark Clarkson 2018

First published in 2018 by

Hodder Education,
An Hachette UK company
Carmelite House
50 Victoria Embankment
London EC4Y 0DZ
www.hoddereducation.co.uk

Impression number   10 9 8 7 6 5

Year   2022

Typeset by Aptara, India

Printed in the UK

A catalogue record for this title is available from the British Library.

HODDER EDUCATION
t: 01235 827827
e: education@hachette.co.uk
w: hoddereducation.co.uk

ISBN 978-1-5104-3702-9

MIX
Paper | Supporting responsible forestry
FSC™ C104740